BRAIN GAMES!
Math Puzzle Books for Adults

TRIANGLE EDITION 2

Smarter Activity Books

INSTRUCTIONS:

1. The top line is filled with numbers. Fill in the empty squares.
2. Take the two numbers of the squares above an empty square.
3. Add these two numbers and write the answer in the empty square.
4. Continue until you reach the bottom square.

example:

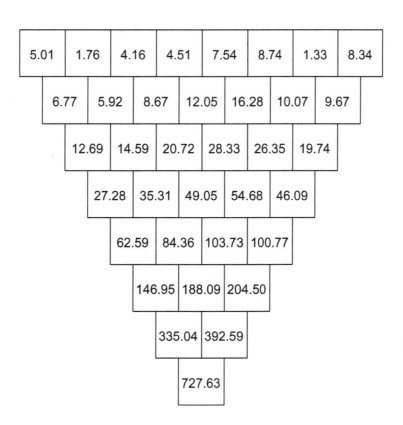

adding

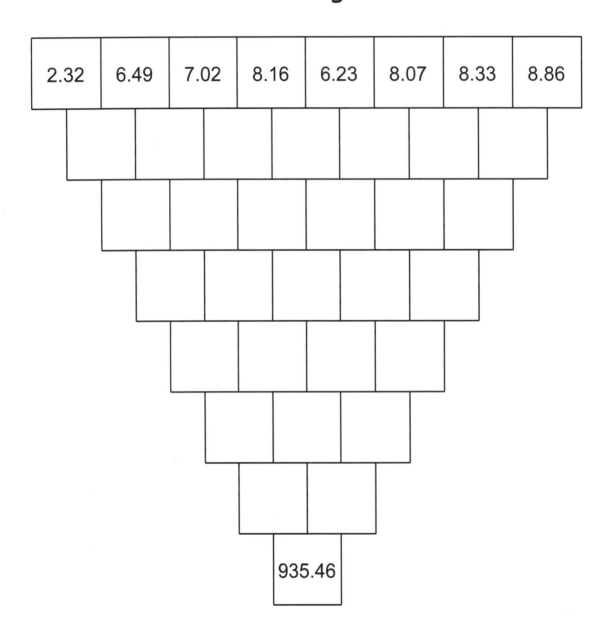

| 2.32 | 6.49 | 7.02 | 8.16 | 6.23 | 8.07 | 8.33 | 8.86 |

adding

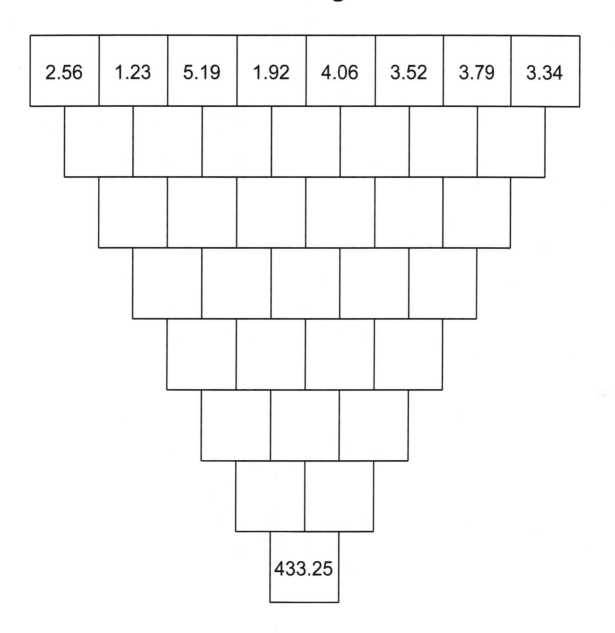

| 2.56 | 1.23 | 5.19 | 1.92 | 4.06 | 3.52 | 3.79 | 3.34 |

433.25

adding

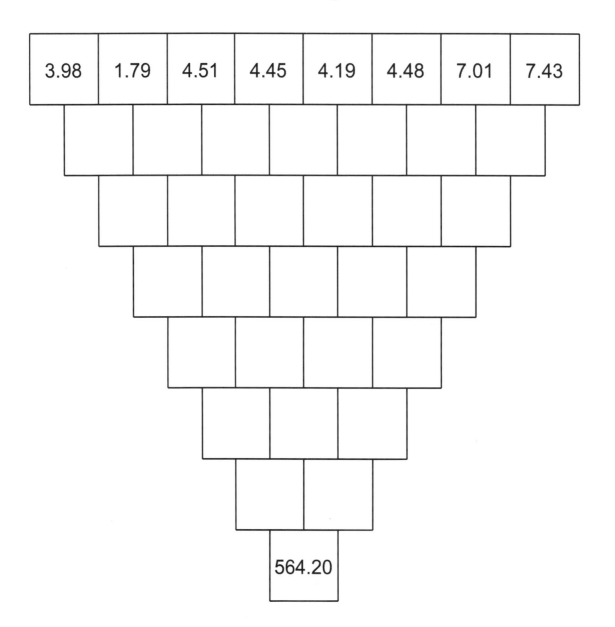

adding

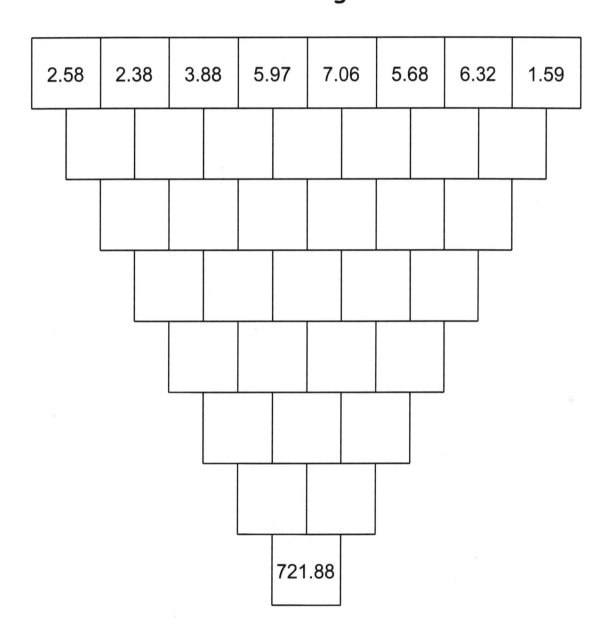

| 2.58 | 2.38 | 3.88 | 5.97 | 7.06 | 5.68 | 6.32 | 1.59 |

721.88

adding

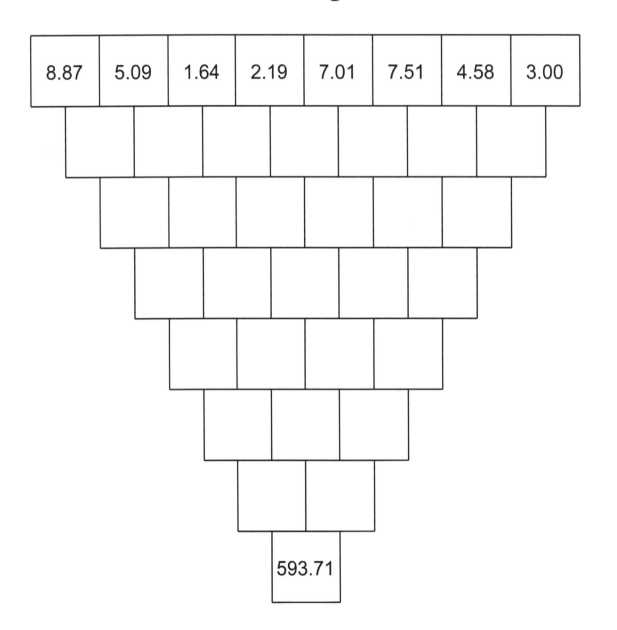

adding

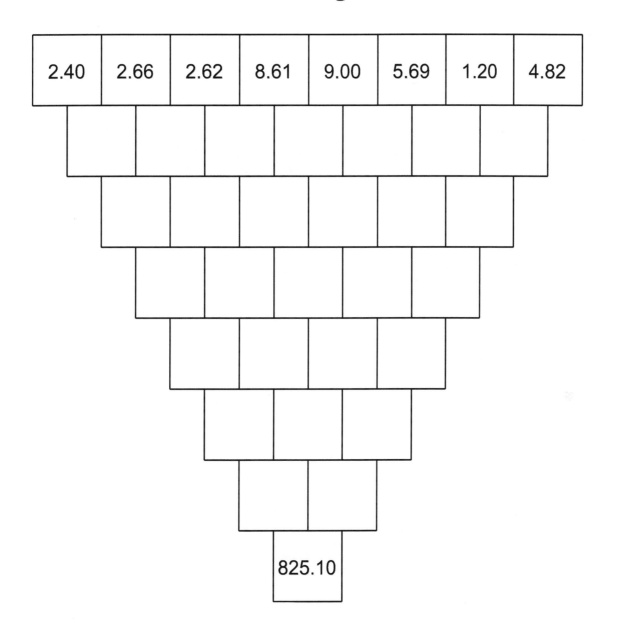

| 2.40 | 2.66 | 2.62 | 8.61 | 9.00 | 5.69 | 1.20 | 4.82 |

825.10

adding

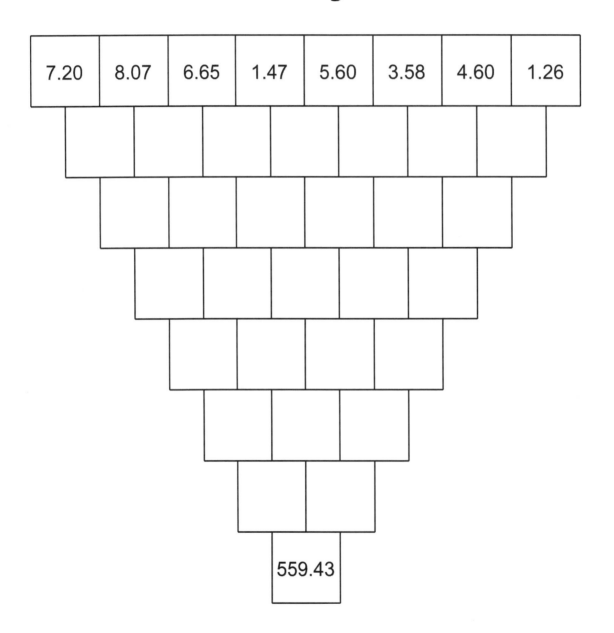

| 7.20 | 8.07 | 6.65 | 1.47 | 5.60 | 3.58 | 4.60 | 1.26 |

559.43

adding

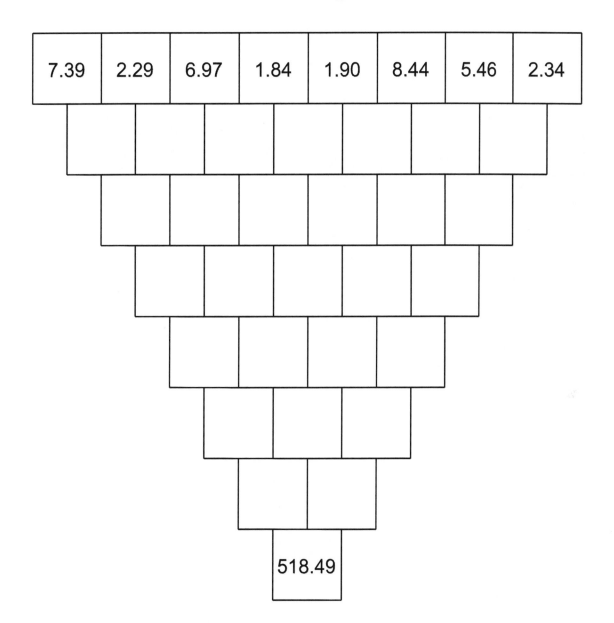

| 7.39 | 2.29 | 6.97 | 1.84 | 1.90 | 8.44 | 5.46 | 2.34 |

518.49

adding

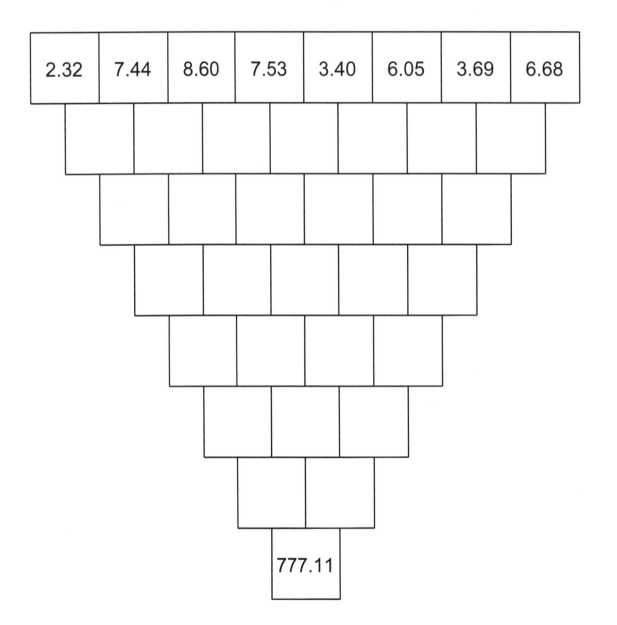

2.32	7.44	8.60	7.53	3.40	6.05	3.69	6.68

777.11

adding

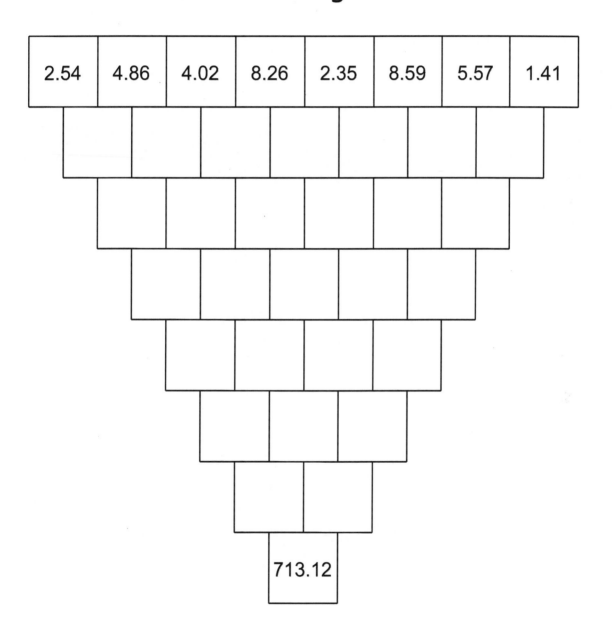

2.54	4.86	4.02	8.26	2.35	8.59	5.57	1.41

713.12

adding

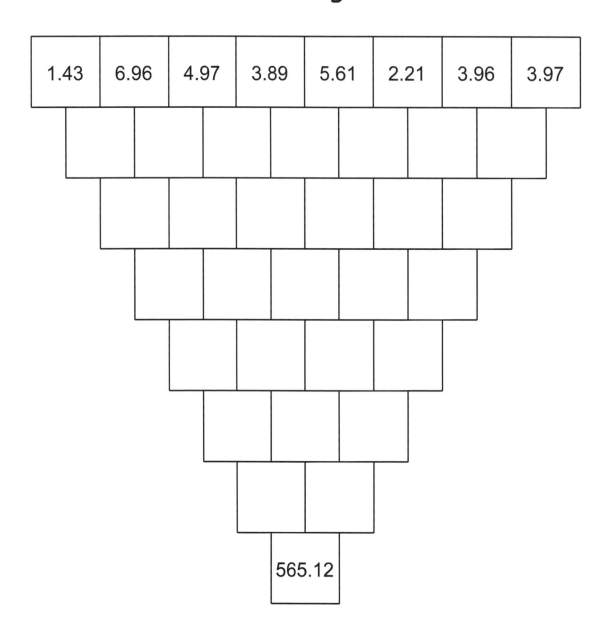

1.43	6.96	4.97	3.89	5.61	2.21	3.96	3.97

565.12

adding

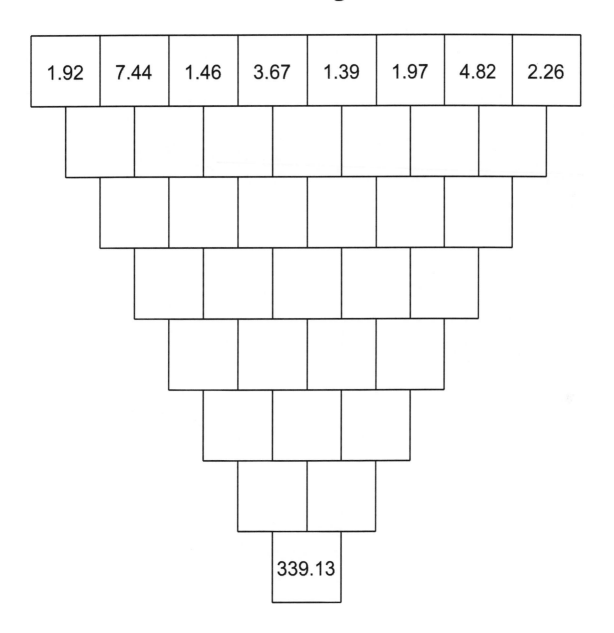

1.92	7.44	1.46	3.67	1.39	1.97	4.82	2.26

339.13

adding

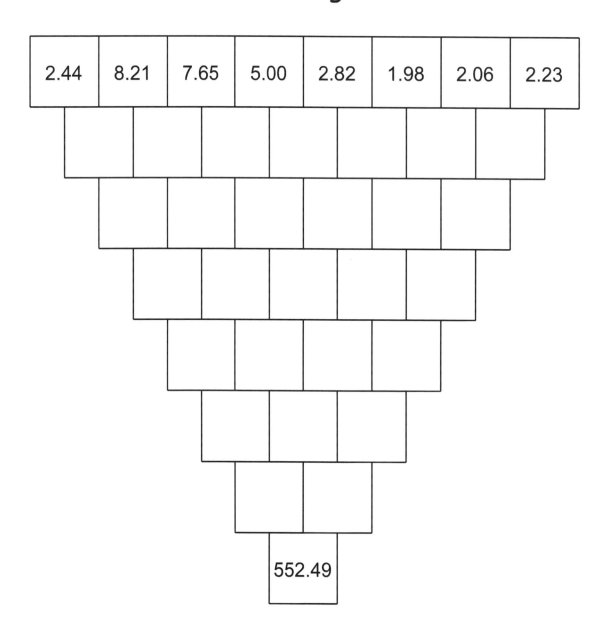

| 2.44 | 8.21 | 7.65 | 5.00 | 2.82 | 1.98 | 2.06 | 2.23 |

552.49

adding

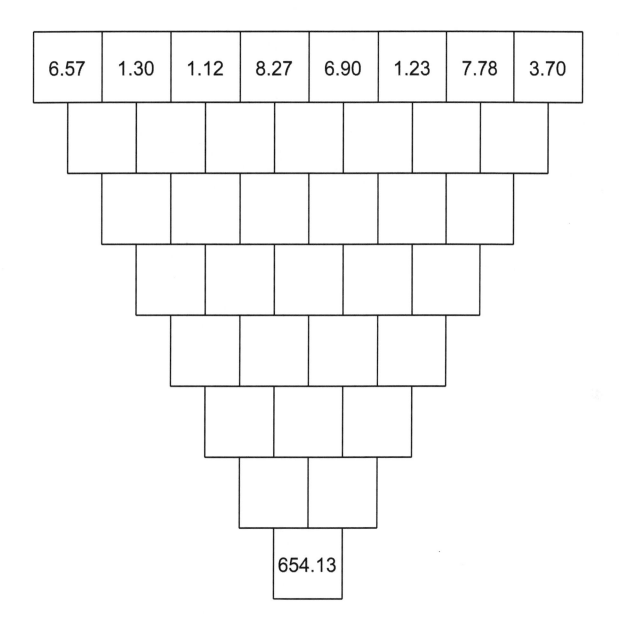

| 6.57 | 1.30 | 1.12 | 8.27 | 6.90 | 1.23 | 7.78 | 3.70 |

654.13

adding

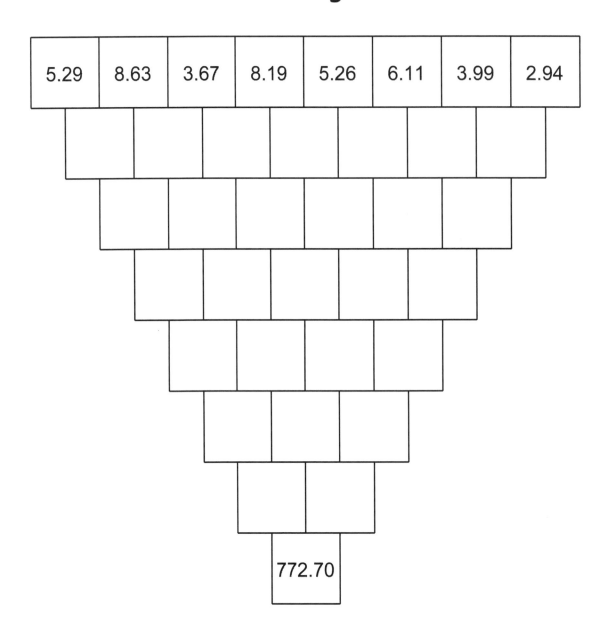

adding

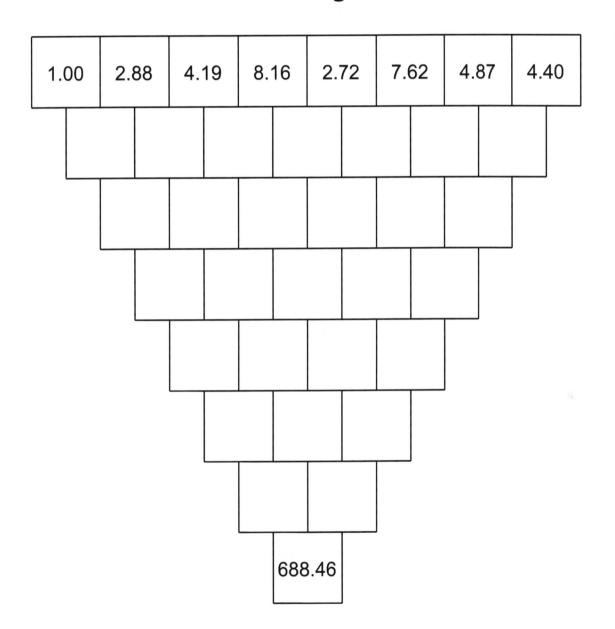

| 1.00 | 2.88 | 4.19 | 8.16 | 2.72 | 7.62 | 4.87 | 4.40 |

688.46

adding

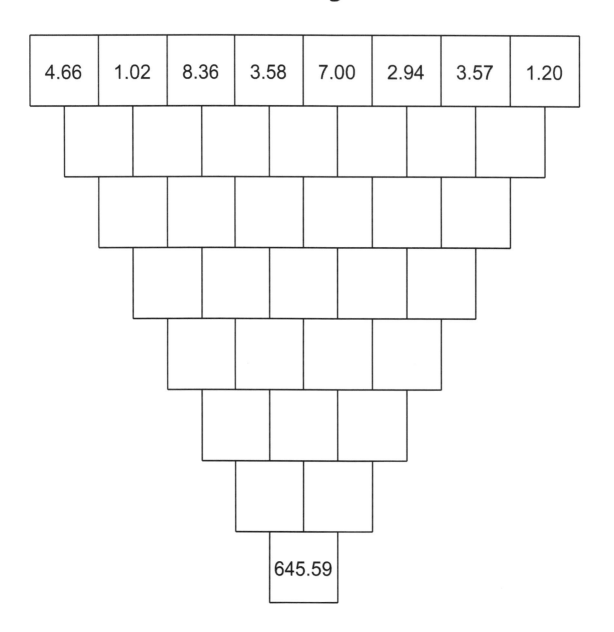

adding

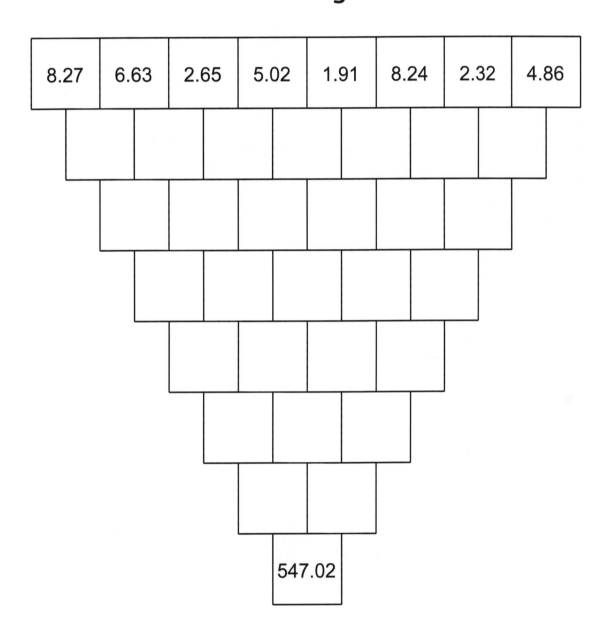

8.27 6.63 2.65 5.02 1.91 8.24 2.32 4.86

547.02

adding

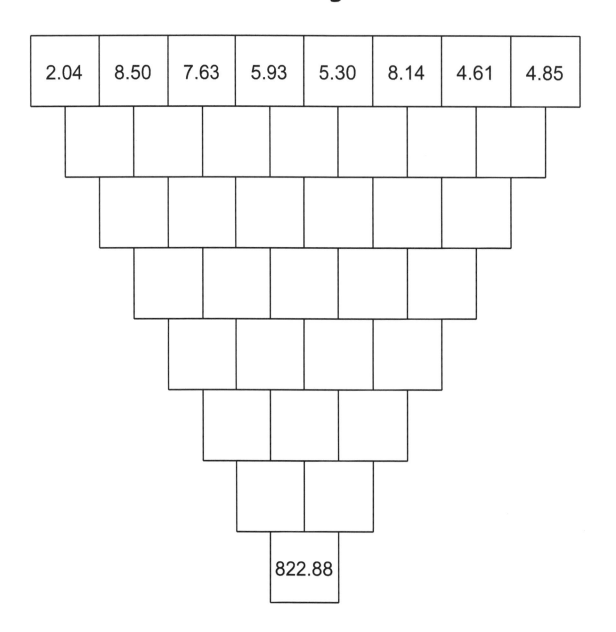

2.04	8.50	7.63	5.93	5.30	8.14	4.61	4.85

822.88

adding

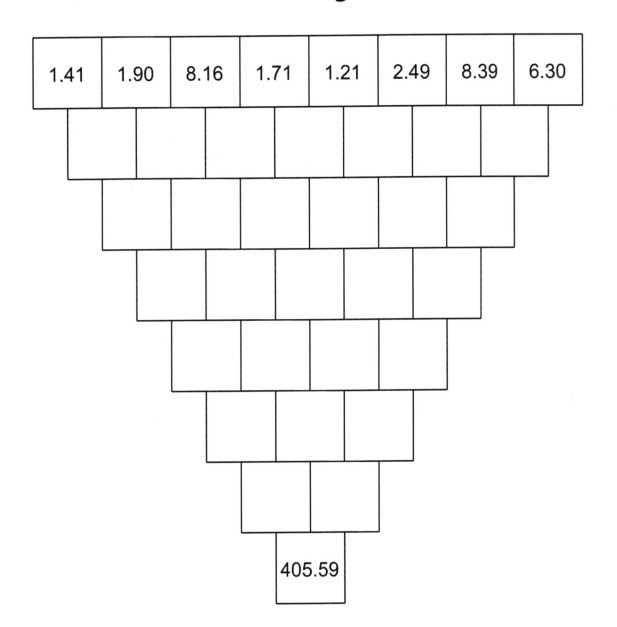

1.41	1.90	8.16	1.71	1.21	2.49	8.39	6.30

405.59

adding

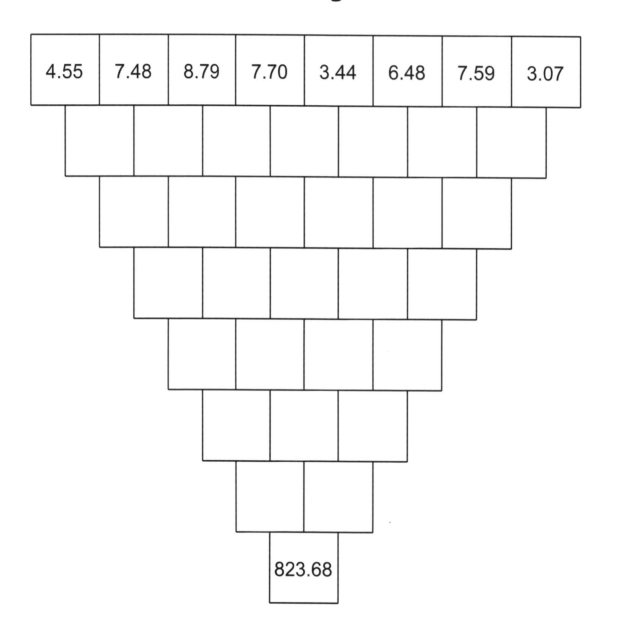

4.55	7.48	8.79	7.70	3.44	6.48	7.59	3.07

823.68

adding

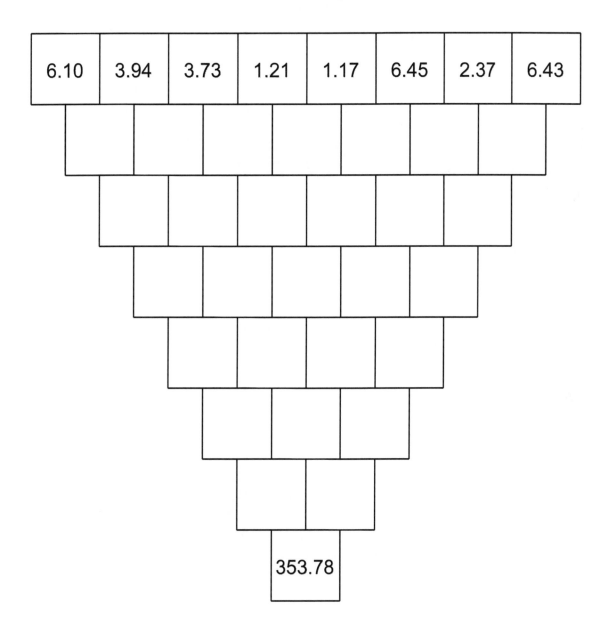

| 6.10 | 3.94 | 3.73 | 1.21 | 1.17 | 6.45 | 2.37 | 6.43 |

353.78

adding

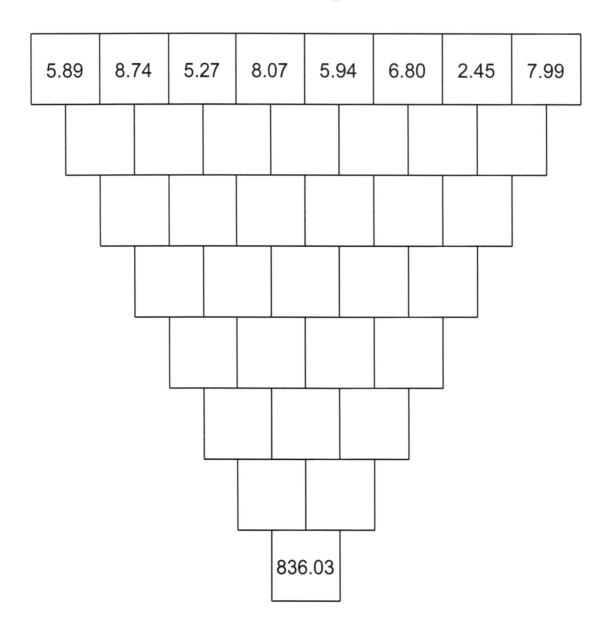

| 5.89 | 8.74 | 5.27 | 8.07 | 5.94 | 6.80 | 2.45 | 7.99 |

836.03

adding

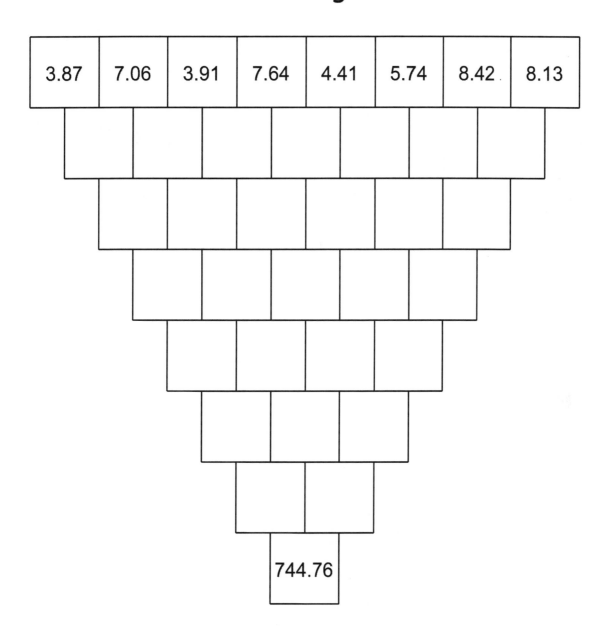

| 3.87 | 7.06 | 3.91 | 7.64 | 4.41 | 5.74 | 8.42 | 8.13 |

adding

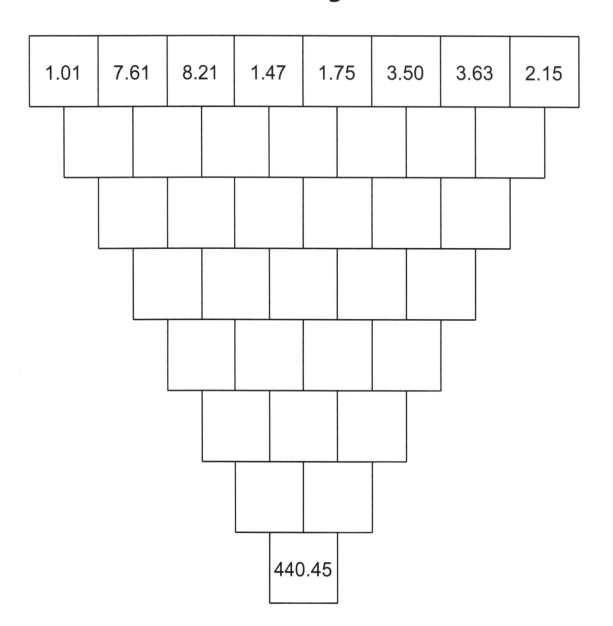

| 1.01 | 7.61 | 8.21 | 1.47 | 1.75 | 3.50 | 3.63 | 2.15 |

440.45

adding

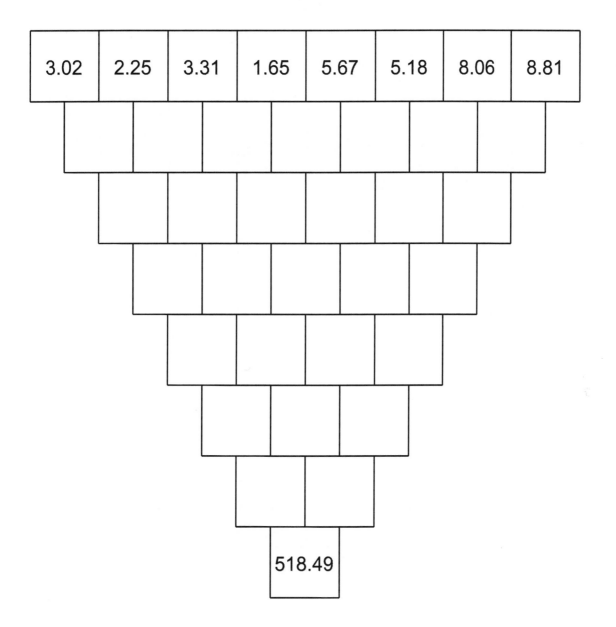

| 3.02 | 2.25 | 3.31 | 1.65 | 5.67 | 5.18 | 8.06 | 8.81 |

518.49

adding

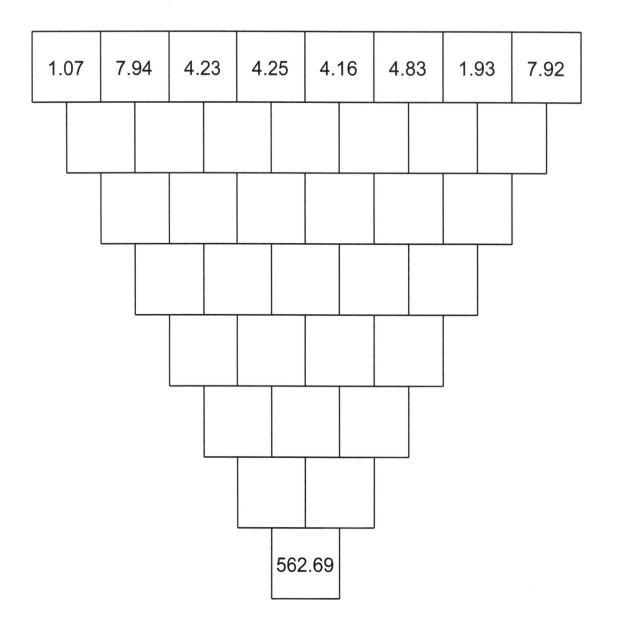

adding

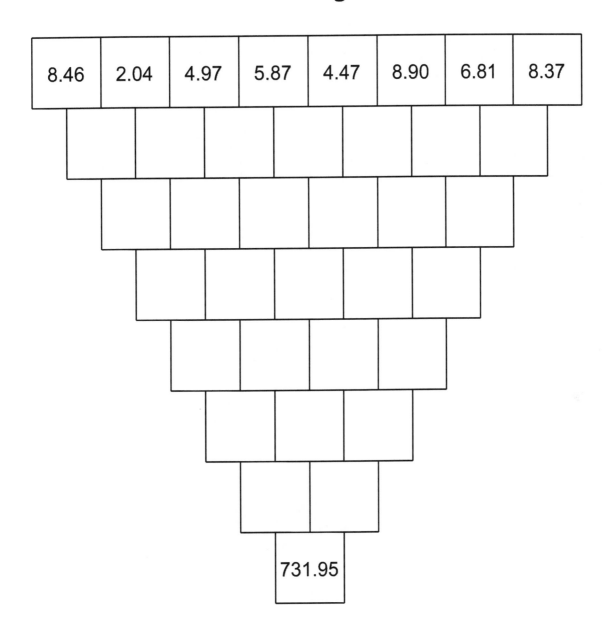

| 8.46 | 2.04 | 4.97 | 5.87 | 4.47 | 8.90 | 6.81 | 8.37 |

731.95

adding

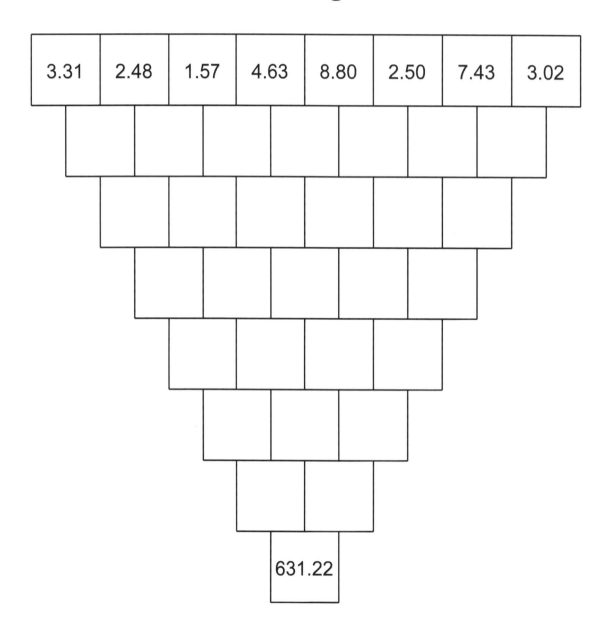

adding

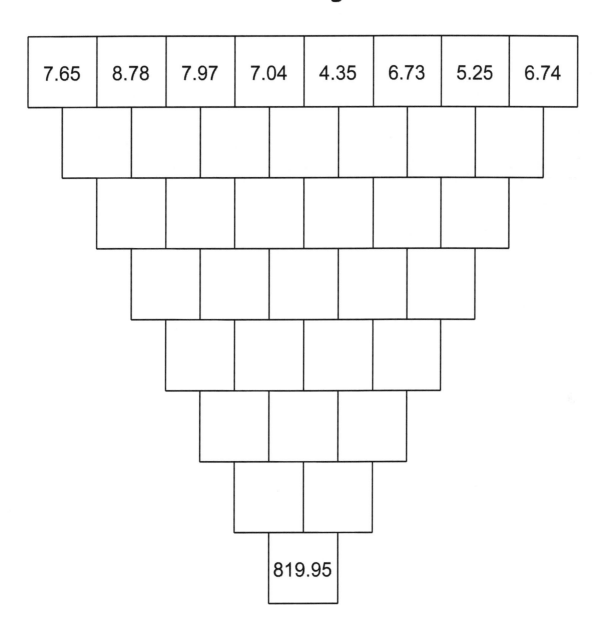

7.65	8.78	7.97	7.04	4.35	6.73	5.25	6.74

819.95

adding

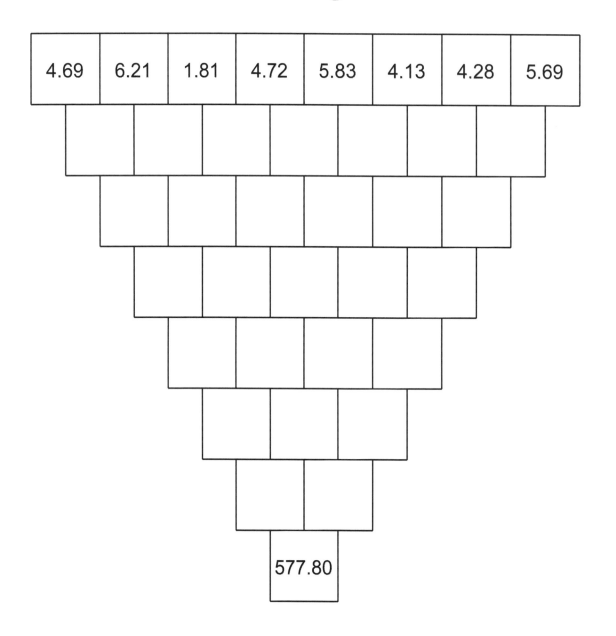

adding

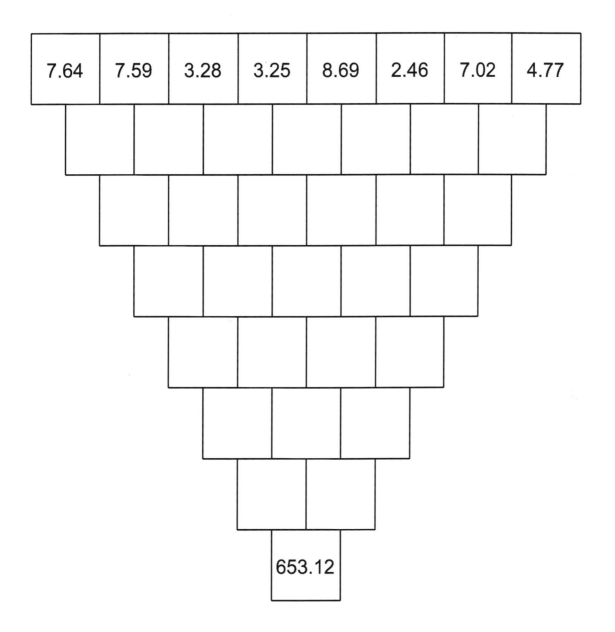

| 7.64 | 7.59 | 3.28 | 3.25 | 8.69 | 2.46 | 7.02 | 4.77 |

653.12

adding

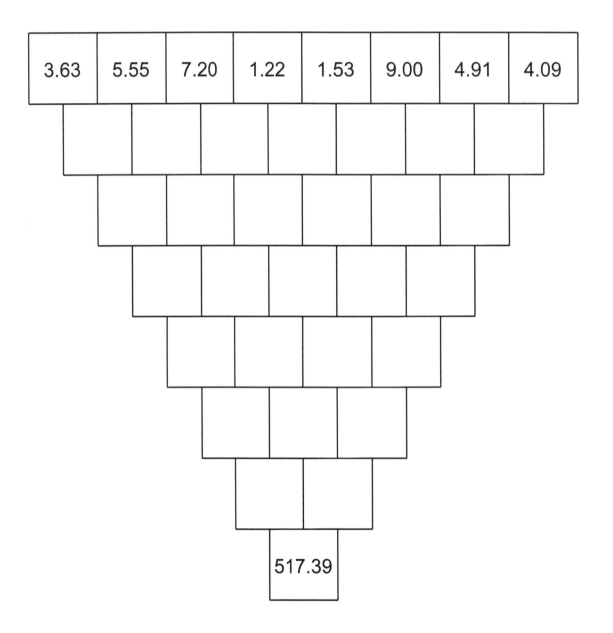

| 3.63 | 5.55 | 7.20 | 1.22 | 1.53 | 9.00 | 4.91 | 4.09 |

517.39

adding

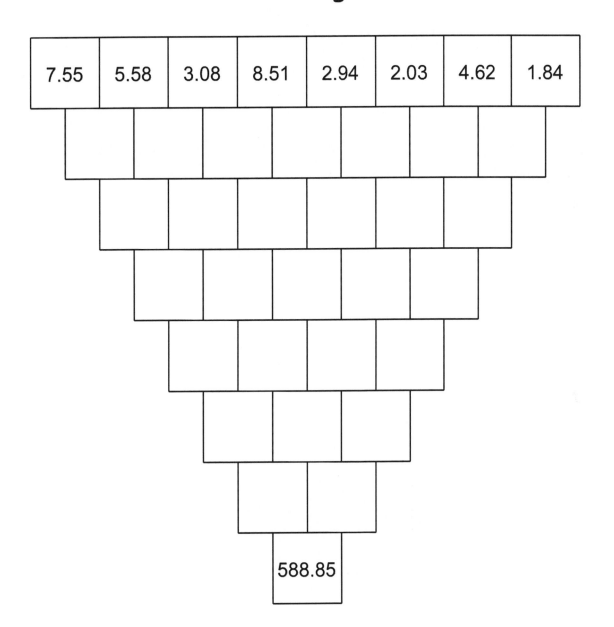

| 7.55 | 5.58 | 3.08 | 8.51 | 2.94 | 2.03 | 4.62 | 1.84 |

588.85

adding

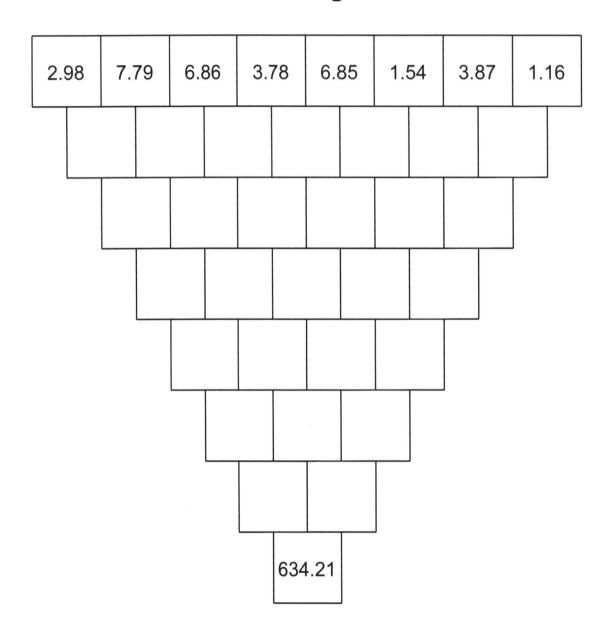

| 2.98 | 7.79 | 6.86 | 3.78 | 6.85 | 1.54 | 3.87 | 1.16 |

634.21

adding

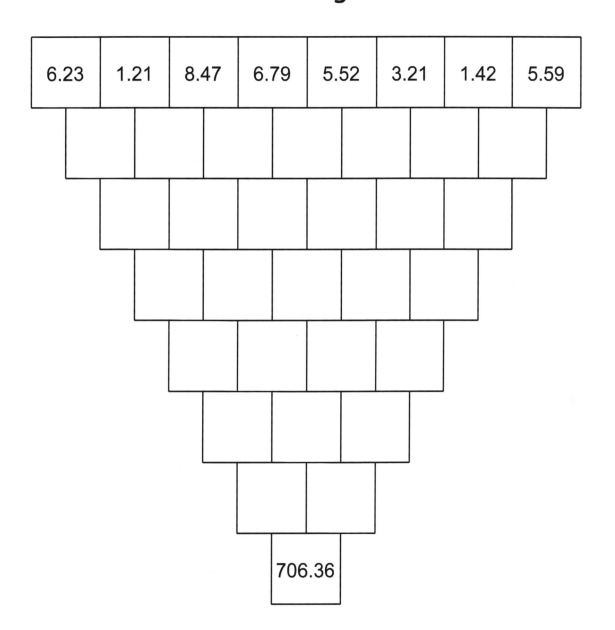

| 6.23 | 1.21 | 8.47 | 6.79 | 5.52 | 3.21 | 1.42 | 5.59 |

adding

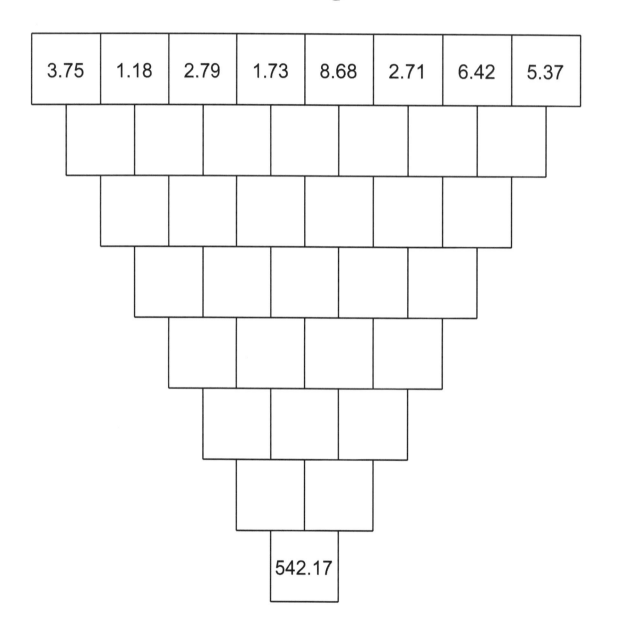

| 3.75 | 1.18 | 2.79 | 1.73 | 8.68 | 2.71 | 6.42 | 5.37 |

542.17

adding

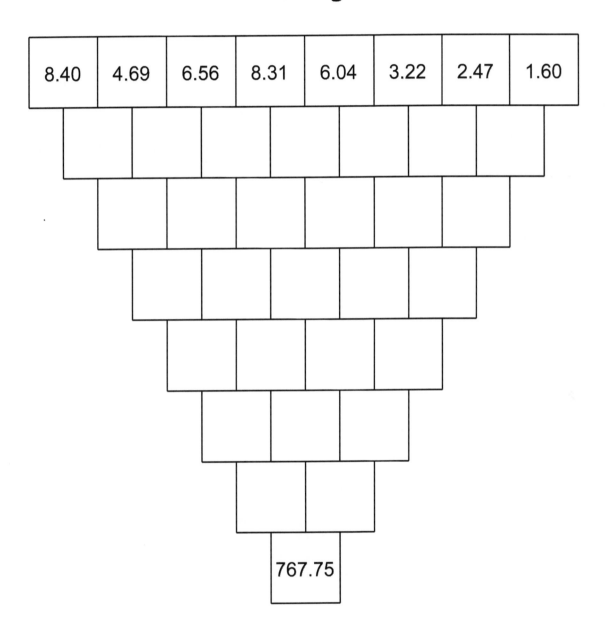

| 8.40 | 4.69 | 6.56 | 8.31 | 6.04 | 3.22 | 2.47 | 1.60 |

767.75

adding

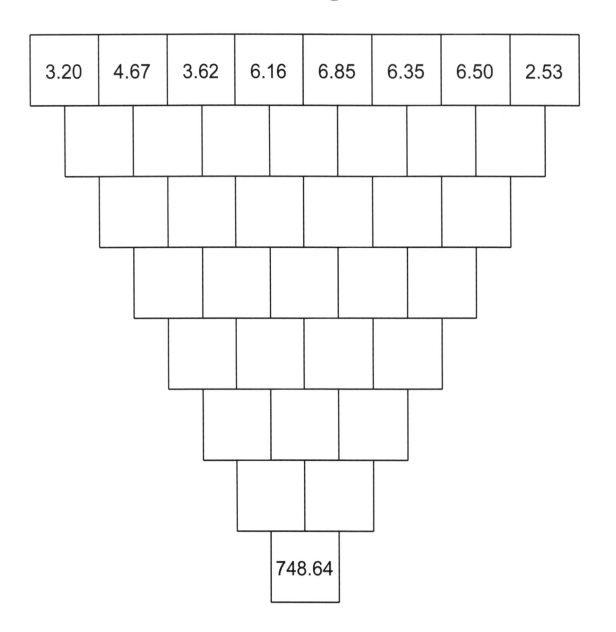

3.20	4.67	3.62	6.16	6.85	6.35	6.50	2.53

748.64

adding

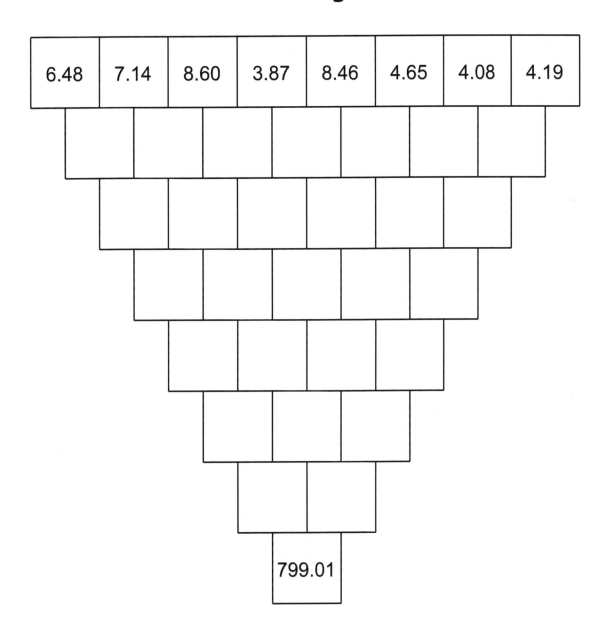

| 6.48 | 7.14 | 8.60 | 3.87 | 8.46 | 4.65 | 4.08 | 4.19 |

799.01

adding

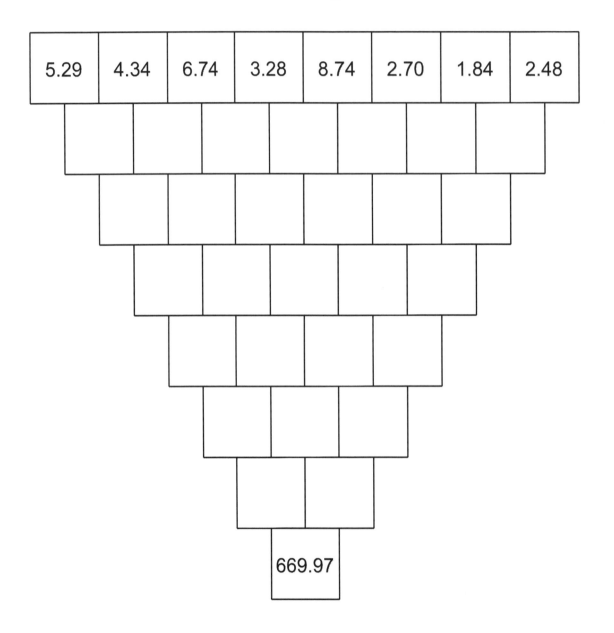

5.29	4.34	6.74	3.28	8.74	2.70	1.84	2.48

669.97

adding

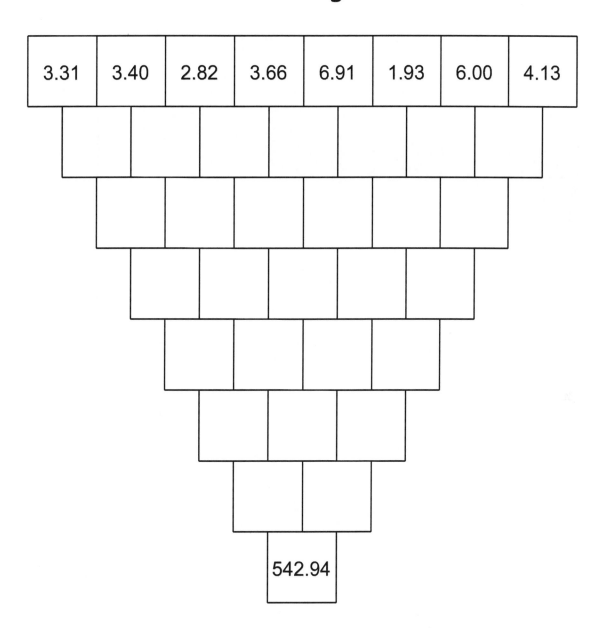

| 3.31 | 3.40 | 2.82 | 3.66 | 6.91 | 1.93 | 6.00 | 4.13 |

542.94

adding

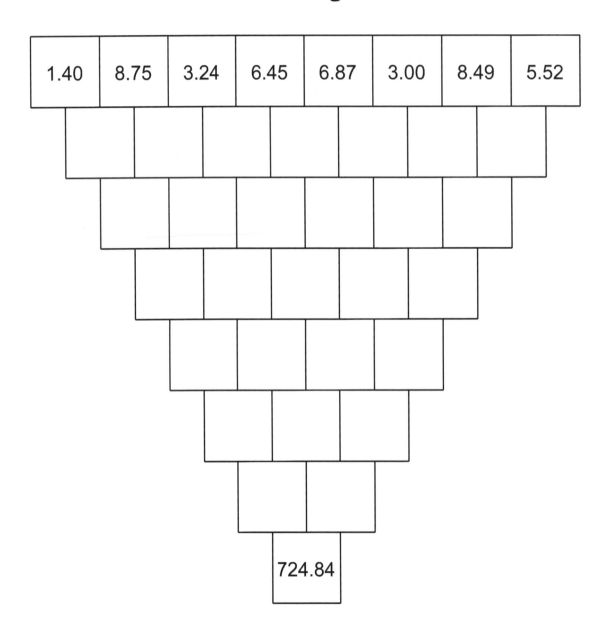

| 1.40 | 8.75 | 3.24 | 6.45 | 6.87 | 3.00 | 8.49 | 5.52 |

724.84

adding

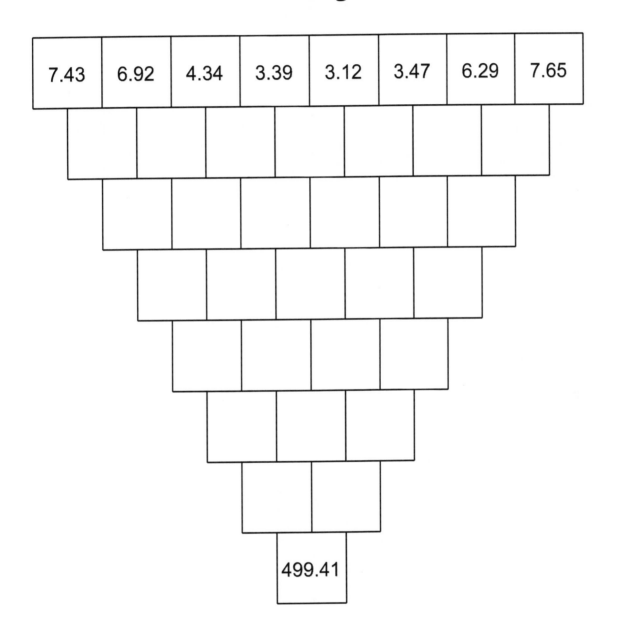

| 7.43 | 6.92 | 4.34 | 3.39 | 3.12 | 3.47 | 6.29 | 7.65 |

499.41

adding

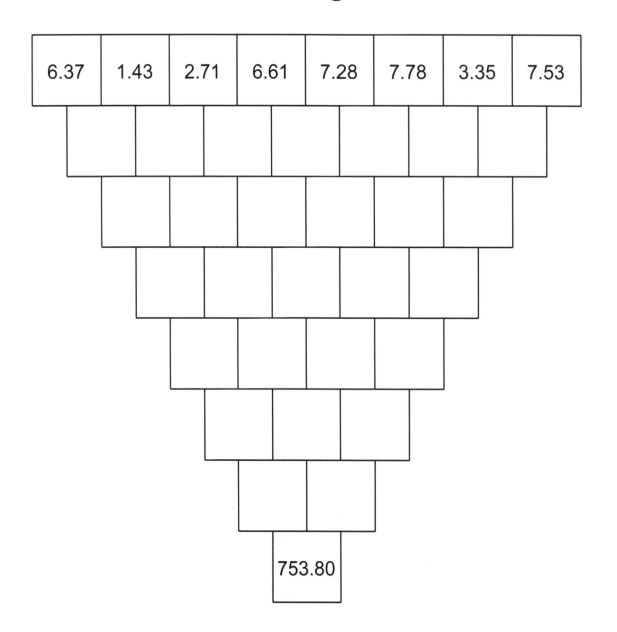

| 6.37 | 1.43 | 2.71 | 6.61 | 7.28 | 7.78 | 3.35 | 7.53 |

753.80

adding

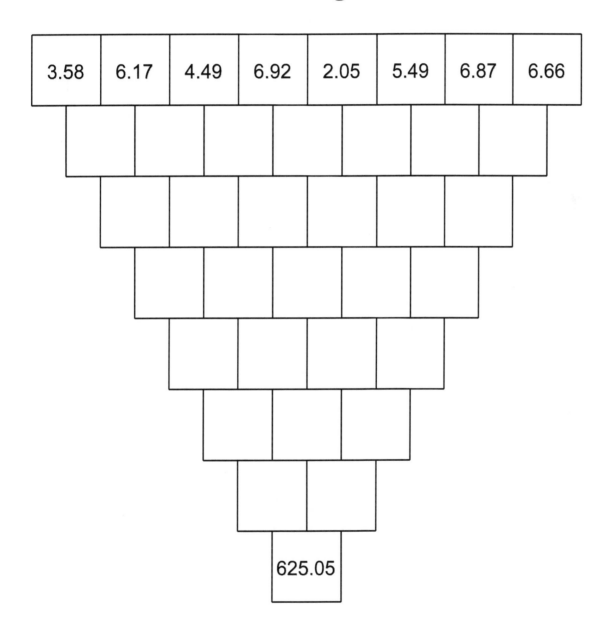

| 3.58 | 6.17 | 4.49 | 6.92 | 2.05 | 5.49 | 6.87 | 6.66 |

625.05

adding

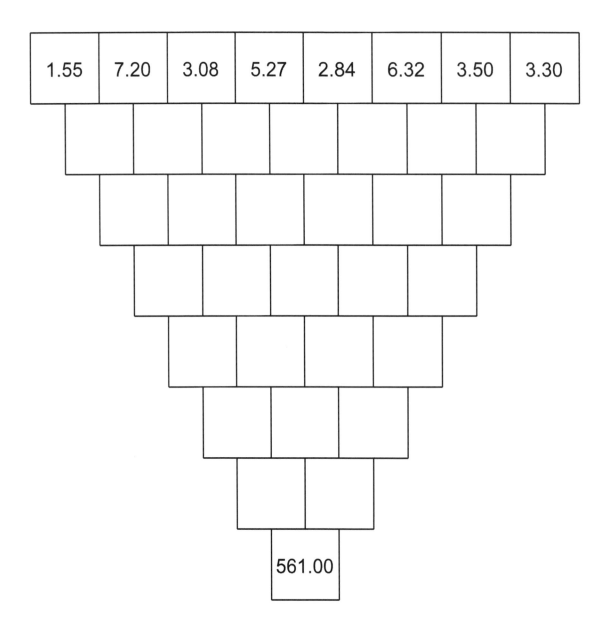

1.55	7.20	3.08	5.27	2.84	6.32	3.50	3.30

561.00

ANSWER

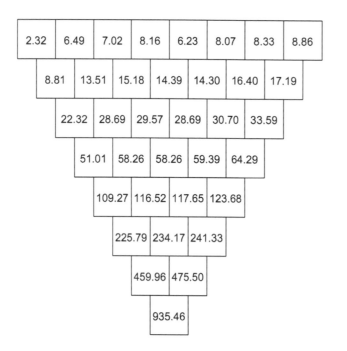

Activity # 1

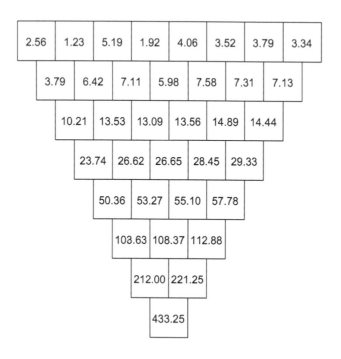

Activity # 2

Activity # 3

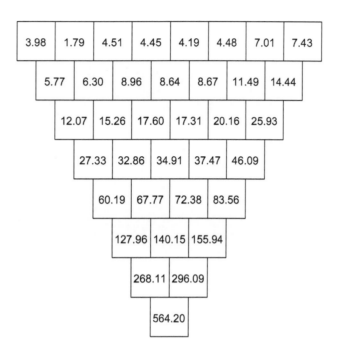

Activity # 4

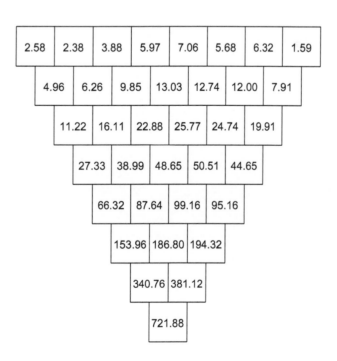

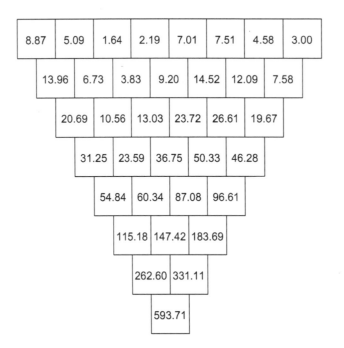

Activity # 5

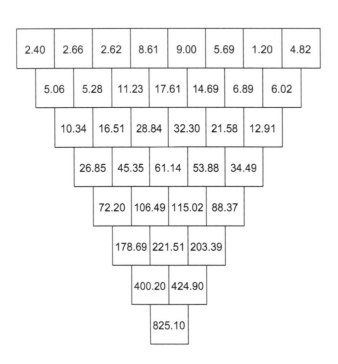

Activity # 6

Activity # 7

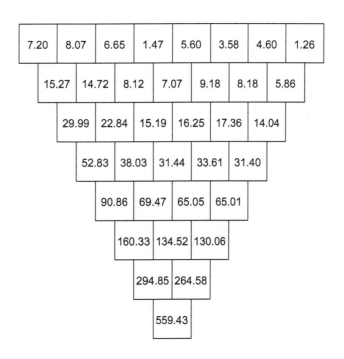

7.20	8.07	6.65	1.47	5.60	3.58	4.60	1.26

15.27	14.72	8.12	7.07	9.18	8.18	5.86

29.99	22.84	15.19	16.25	17.36	14.04

52.83	38.03	31.44	33.61	31.40

90.86	69.47	65.05	65.01

160.33	134.52	130.06

294.85	264.58

559.43

Activity # 8

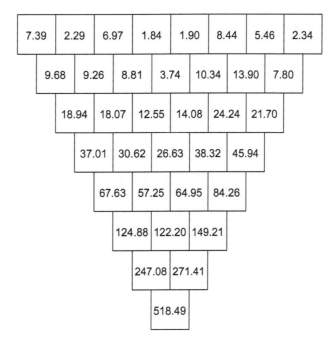

7.39	2.29	6.97	1.84	1.90	8.44	5.46	2.34

9.68	9.26	8.81	3.74	10.34	13.90	7.80

18.94	18.07	12.55	14.08	24.24	21.70

37.01	30.62	26.63	38.32	45.94

67.63	57.25	64.95	84.26

124.88	122.20	149.21

247.08	271.41

518.49

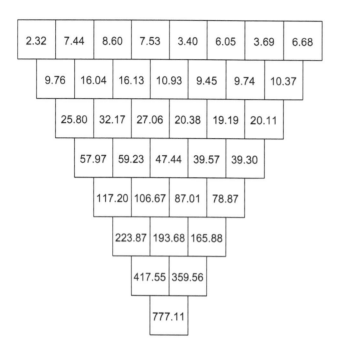

Activity # 9

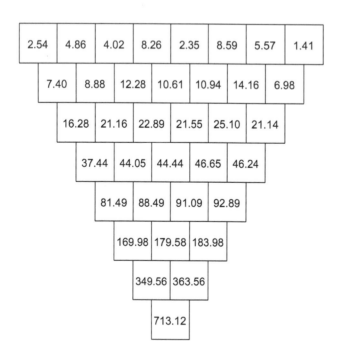

Activity # 10

Activity # 11

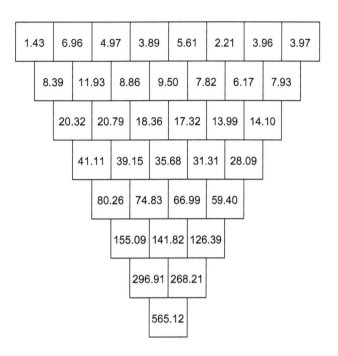

1.43	6.96	4.97	3.89	5.61	2.21	3.96	3.97

8.39	11.93	8.86	9.50	7.82	6.17	7.93

20.32	20.79	18.36	17.32	13.99	14.10

41.11	39.15	35.68	31.31	28.09

80.26	74.83	66.99	59.40

155.09	141.82	126.39

296.91	268.21

565.12

Activity # 12

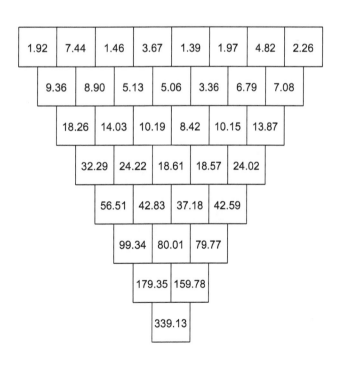

1.92	7.44	1.46	3.67	1.39	1.97	4.82	2.26

9.36	8.90	5.13	5.06	3.36	6.79	7.08

18.26	14.03	10.19	8.42	10.15	13.87

32.29	24.22	18.61	18.57	24.02

56.51	42.83	37.18	42.59

99.34	80.01	79.77

179.35	159.78

339.13

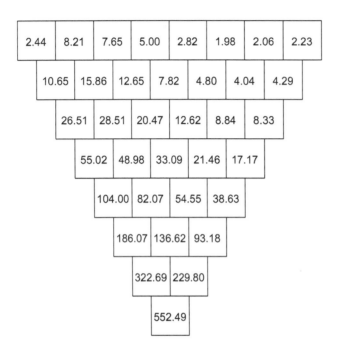

Activity # 13

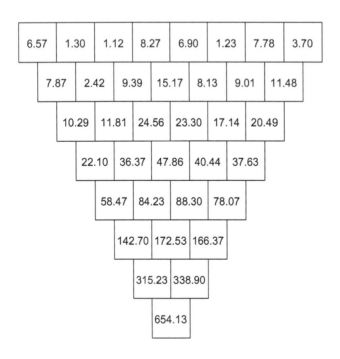

Activity # 14

Activity # 15

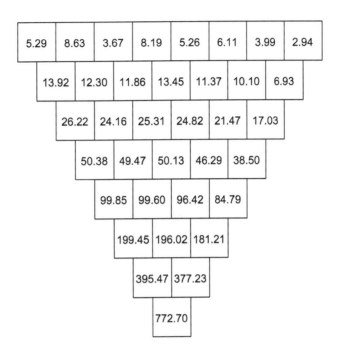

Activity # 16

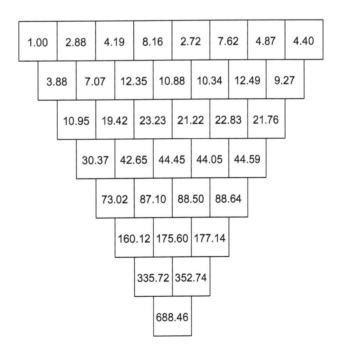

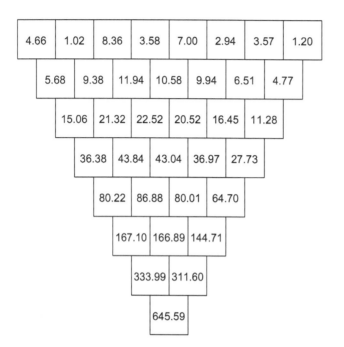

Activity # 17

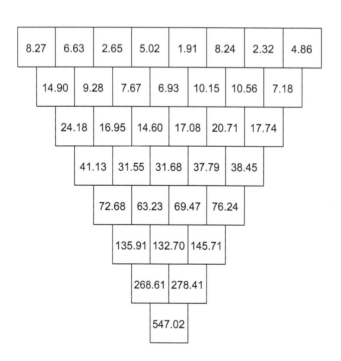

Activity # 18

Activity # 19

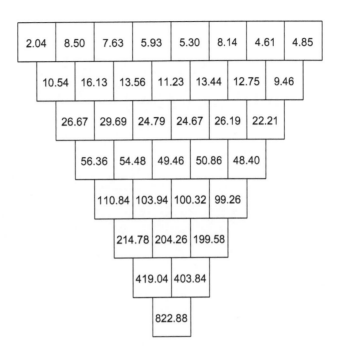

| 2.04 | 8.50 | 7.63 | 5.93 | 5.30 | 8.14 | 4.61 | 4.85 |

| 10.54 | 16.13 | 13.56 | 11.23 | 13.44 | 12.75 | 9.46 |

| 26.67 | 29.69 | 24.79 | 24.67 | 26.19 | 22.21 |

| 56.36 | 54.48 | 49.46 | 50.86 | 48.40 |

| 110.84 | 103.94 | 100.32 | 99.26 |

| 214.78 | 204.26 | 199.58 |

| 419.04 | 403.84 |

| 822.88 |

Activity # 20

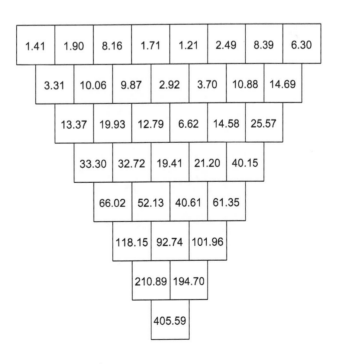

| 1.41 | 1.90 | 8.16 | 1.71 | 1.21 | 2.49 | 8.39 | 6.30 |

| 3.31 | 10.06 | 9.87 | 2.92 | 3.70 | 10.88 | 14.69 |

| 13.37 | 19.93 | 12.79 | 6.62 | 14.58 | 25.57 |

| 33.30 | 32.72 | 19.41 | 21.20 | 40.15 |

| 66.02 | 52.13 | 40.61 | 61.35 |

| 118.15 | 92.74 | 101.96 |

| 210.89 | 194.70 |

| 405.59 |

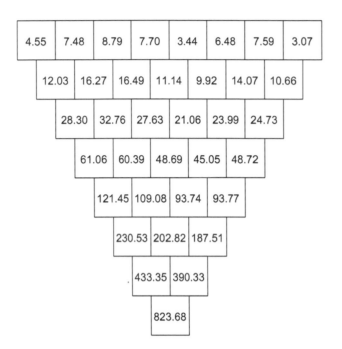

Activity # 21

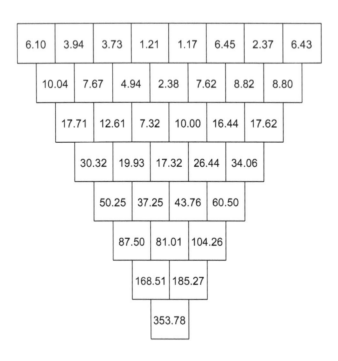

Activity # 22

Activity # 23

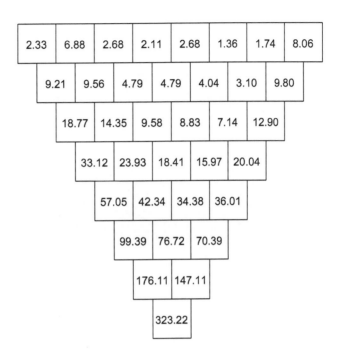

2.33	6.88	2.68	2.11	2.68	1.36	1.74	8.06

9.21	9.56	4.79	4.79	4.04	3.10	9.80

18.77	14.35	9.58	8.83	7.14	12.90

33.12	23.93	18.41	15.97	20.04

57.05	42.34	34.38	36.01

99.39	76.72	70.39

176.11	147.11

323.22

Activity # 24

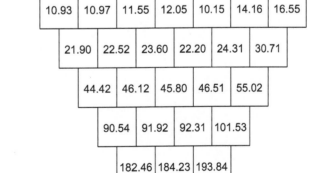

3.87	7.06	3.91	7.64	4.41	5.74	8.42	8.13

10.93	10.97	11.55	12.05	10.15	14.16	16.55

21.90	22.52	23.60	22.20	24.31	30.71

44.42	46.12	45.80	46.51	55.02

90.54	91.92	92.31	101.53

182.46	184.23	193.84

366.69	378.07

744.76

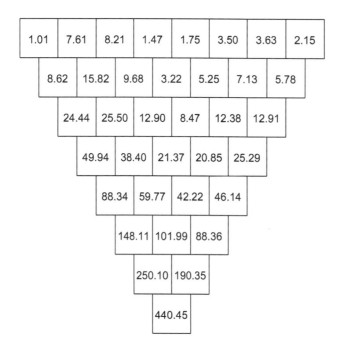

Activity # 25

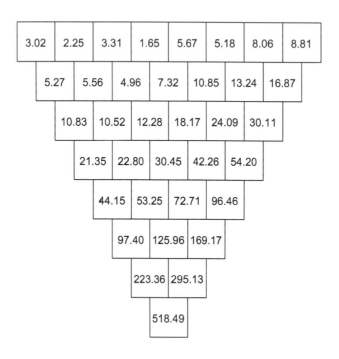

Activity # 26

Activity # 27

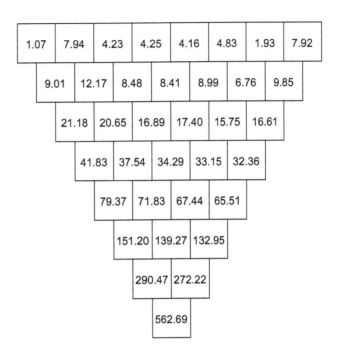

Activity # 28

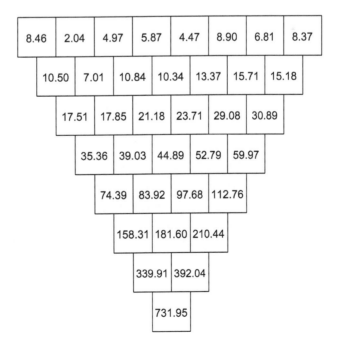

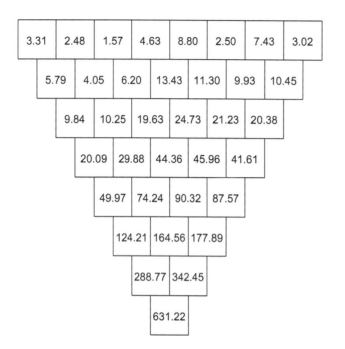

Activity # 29

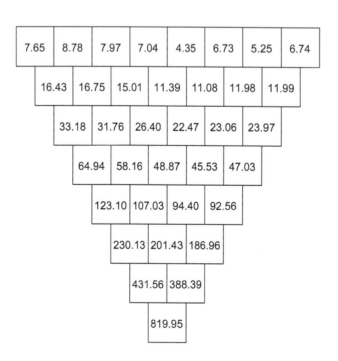

Activity # 30

Activity # 31

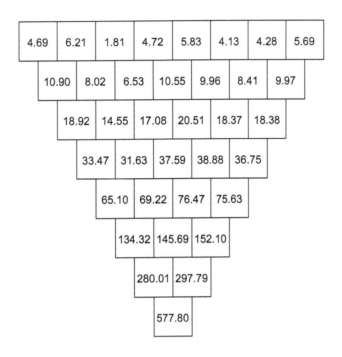

Activity # 32

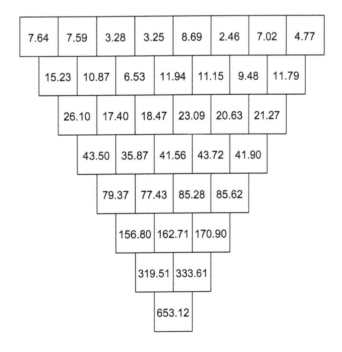

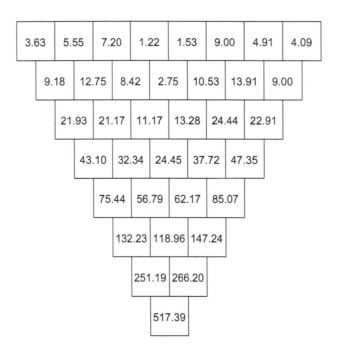

Activity # 33

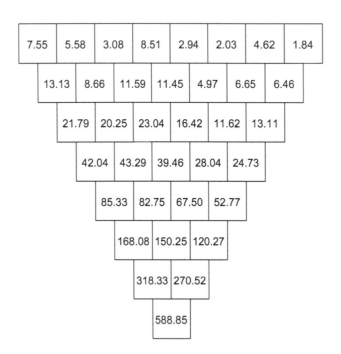

Activity # 34

Activity # 35

2.98	7.79	6.86	3.78	6.85	1.54	3.87	1.16

10.77	14.65	10.64	10.63	8.39	5.41	5.03

25.42	25.29	21.27	19.02	13.80	10.44

50.71	46.56	40.29	32.82	24.24

97.27	86.85	73.11	57.06

184.12	159.96	130.17

344.08	290.13

634.21

Activity # 36

6.23	1.21	8.47	6.79	5.52	3.21	1.42	5.59

7.44	9.68	15.26	12.31	8.73	4.63	7.01

17.12	24.94	27.57	21.04	13.36	11.64

42.06	52.51	48.61	34.40	25.00

94.57	101.12	83.01	59.40

195.69	184.13	142.41

379.82	326.54

706.36

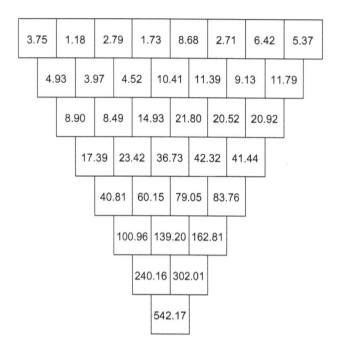

Activity # 37

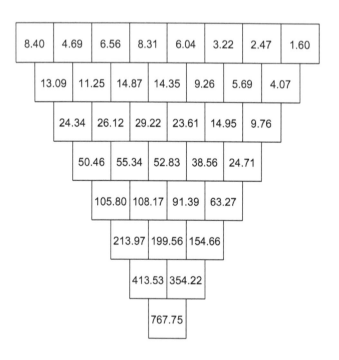

Activity # 38

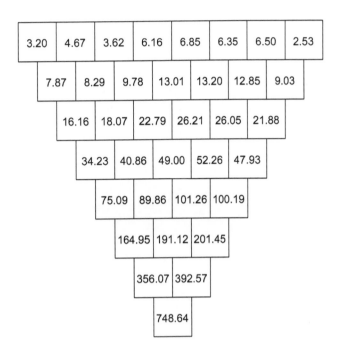

Activity # 39

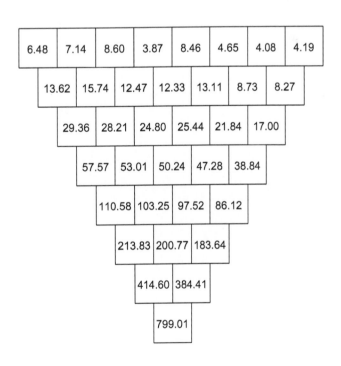

Activity # 40

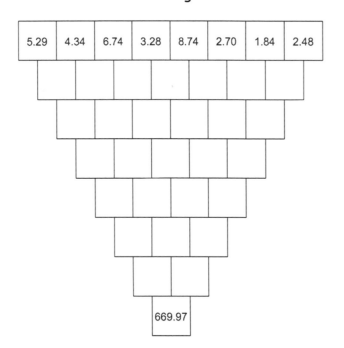

Activity # 41

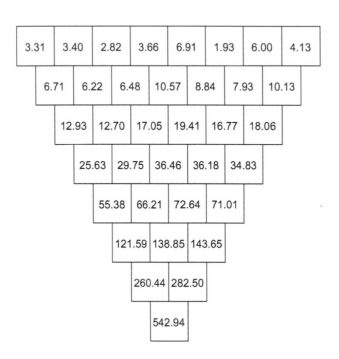

Activity # 42

Activity # 43

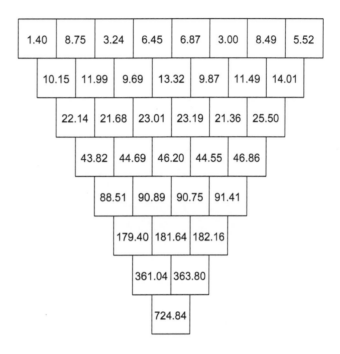

1.40	8.75	3.24	6.45	6.87	3.00	8.49	5.52
	10.15	11.99	9.69	13.32	9.87	11.49	14.01
		22.14	21.68	23.01	23.19	21.36	25.50
			43.82	44.69	46.20	44.55	46.86
				88.51	90.89	90.75	91.41
					179.40	181.64	182.16
						361.04	363.80
							724.84

Activity #44

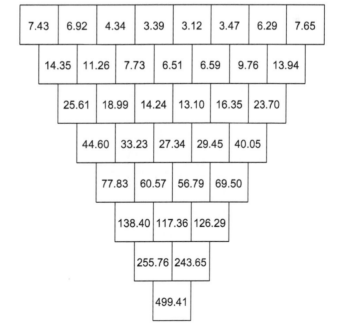

7.43	6.92	4.34	3.39	3.12	3.47	6.29	7.65
	14.35	11.26	7.73	6.51	6.59	9.76	13.94
		25.61	18.99	14.24	13.10	16.35	23.70
			44.60	33.23	27.34	29.45	40.05
				77.83	60.57	56.79	69.50
					138.40	117.36	126.29
						255.76	243.65
							499.41

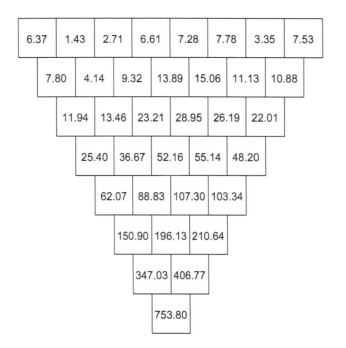

Activity #45

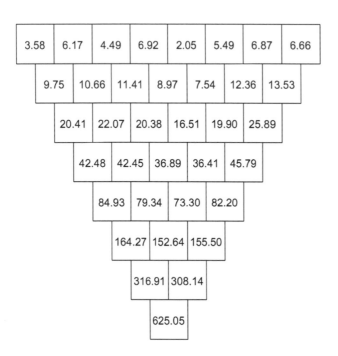

Activity # 46

Activity # 47

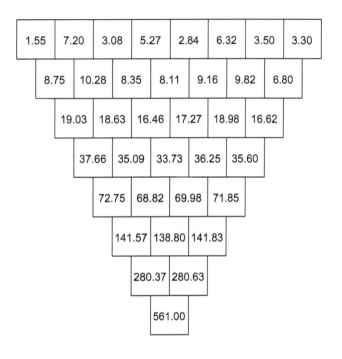

1.55	7.20	3.08	5.27	2.84	6.32	3.50	3.30

8.75	10.28	8.35	8.11	9.16	9.82	6.80

19.03	18.63	16.46	17.27	18.98	16.62

37.66	35.09	33.73	36.25	35.60

72.75	68.82	69.98	71.85

141.57	138.80	141.83

280.37	280.63

561.00

Printed in Great Britain
by Amazon